Blue Cactus Press and our staff work, rest, and reside on traditional territory of the Coast Salish peoples, specifically on land of the Puyallup and Nisqually Tribes. This land was taken under duress via the signing of the Treaty of Medicine Creek in 1854. Since then, it has not been returned to the peoples it belongs to.

Here, we acknowledge that we benefit and profit from our existence in the South Puget Sound on Puyallup and Nisqually land. We respect and honor these peoples and the land. We also understand this acknowledgement does not replace the work of building relationships or trust–it is merely the first step on our path to doing so.

Here, we give thanks. We raise our hands.

PRAISE FOR THIS BOOK

"Coming of age in a world that sells romance as the holy grail of love and gives you hookup apps as a means to accomplish the task, *Confessions of a Modern-Day Kumiho* dares to make sense of whether a meaningful relationship is even possible. Kumiho awakens to the ancient power of being a sexual female creature, exploring the roles of prey and predator, and navigating the emotional toll. Alissa Tu's debut experimental memoir gives voice to the deep hunger for a profound soul connection that lives inside us all."

– Amy Solo, educator and author of *Mothers, Lovers and Roadside Burials*

"Alissa Tu's debut novella, *Confessions of a Modern-Day Kumiho* follows the treacherous, lavender-scented paths of love of the titular Kumiho, a nine-tailed fox with a taste for men's livers and hearts. Tu's paratactical prose creates a labyrinth that seduces readers right into the blood-soaked fangs of her narrator. *Confessions of a Modern-Day Kumiho* is the sensual, gorey, and brilliantly honest look at the hellish landscape of modern romance and the endless road to self-love. To sum it up in the author's own words, 'the tension burned me.'"

– Carlito Espudo

"A memoir of hunger, sex, shifting narration of the self, and the roles in which we cast our partners and prey – this book is so relatable in the ways you don't want to admit in the daylight. Tu deals in the hurtful, fragile moments of mortality between ancient myths and the modern gods of technology, loneliness, and self-love. And, as promised, Tu gives readers a true confession, where the narrative is piecemeal, unsettling, continually escalating, and where connections which appear subjective at first become intricately woven and broken threads throughout."

– Emilie Rommel Shimkus, author, and actress

"Tu unleashes her record of heartbreak and boy drama, all under the guise of a mythic fox who eats men's hearts. But between DM transcripts and Tumblr anons, these aren't the ramblings of a bloodthirsty creature. Rather, a late night goss sesh with a good friend."

– Jonah Barrett, author of *Moss Covered Claws* and director of *The Tetrahedron*

"A unique glimpse into a creative soul, *Confessions of a Modern-Day Kumiho*, traverses a young woman's life in the intimate moments of relationships. Told through myth and metaphor, the moments feel both universal and uniquely personal. As she struggles with the mythos of love, the realities of modern dating, the cultural racism around Asian women, and the objectification of all women, Alissa Tu doesn't shy away from self-reflection but chooses to craft a narrative that is all her own."

– Bethany Maines, award-winning author of *Bulletproof Mascara*

"Watch Out Readers. Check your hearts, livers, kidneys. Make sure they are still there. Alissa Tu is here, ready to devour what most feminists would consider as marginally devoted, demoted, decapitatable. Tu's sassy and supercool work is better than any dating apps! Swipe right into her cloister or jailhouse and soar!"

– Vi Khi Nao, author of *Sheep Machine*, *Umbilical Hospital*, and *A Brief Alphabet of Torture*

"Don't be surprised If you find yourself rooting for the beast who eats the flesh of men who betrayed her. Thoughts of revenge seem reasonable when the object of our affection poisons our self-assurance with lies and indifference. Or maybe we are the ones who betray ourselves through self-delusion, by seeing more than is there in the first place? Alissa Tu's observation that, 'My fatal flaw is letting men who put my heart aside crawl back,' is highly relatable. Results may vary, though I'm certain reading *Confessions of a Modern-Day*

Kumiho will have you looking back at your own dating history, wondering which of those situations might have been improved by a little organ harvesting."

– CK Combs, queer trans writer and community activist, author of an upcoming novel featuring nonbinary youth, ghosts, and an abandoned amusement park

"*Confessions of a Modern-Day Kumiho* is a book about many things: modern dating, the damaging day-to-day impact of being objectified as a woman of color, the resistance of imposed gender norms in favor of the anti-caretaker, the ache to be seen and loved wholly. But above all, I think Tu's debut book is about the risks of vulnerability. Tu doesn't just talk about this, she enacts it—creating a sacred vulnerability with and towards the reader."

– Becca Rae Rose, cross-genre writer and co-founder of *KALEIDOSCOPED MAG*

"This book is utterly compelling: painful in its honesty and searching in its scope, while somehow also brutally funny and fortifying. I am in awe of the nine-tailed fox satiating her hunger through these pages, and you will be, too."

– Jac Jemc, author of *False Bingo* and *Empty Theatre*

"The ancient Egyptians erected entire religious systems to praise the humble house-cat's nine uncanny lives, but let's not get all orientalist about survival. Besides, I would swap my own survival, luxuriate in sacrifice, spread open my body for Alissa Tu's nine hungry tails. *Confessions of a Modern-Day Kumiho* is sex and sass, destructive and indestructible, and if you're worthy, it will delight in you, one organ—one orgasm—at a time. It thirsts."

– Lily Hoàng, author of *Underneath* and *A Bestiary*

"This book is a manifesto of self-knowing through metaphor. In these pages, Tu embraces and wrestles with her identity as a young Vietnamese woman who is both mythological creature and wholly mortal. There's a balance that needs to be discovered for the sake of her own survival: is she her reflection in the eyes of her lovers, or is she her own image? What if she explores all her selves as a nine-tailed fox bent on destruction and love, all at once? The result is a sprint of a narrative that makes us question who each of us has been, is now, and could be. I love this. I screamed *fuck yes!* more than a few times while reading it."

– Abby E. Murray, author of *Hail and Farewell*

> Title pages are duuuumb

> We have to have one.

> You do it then

> Confessions of a Modern-Day Kumiho by Alissa Tu

Today 1:03 PM

> ugh. can we at least put a cute photo on it?

CONTENT WARNING

This book contains graphic content that might be upsetting to some, including highly sexualized language and mentions of self-harm and non-consensual sex.

THE SEVERED RIBS

Kumiho: A nine-tailed fox who transforms into a beautiful woman to seduce men and eat their livers or hearts.

I gave a boy my writing heart, a book that became a foundation to my writing. This book housed my soul, too. Three months later, he told me he didn't read it. He put my heart on a shelf and let it collect cobwebs. It was only a paperweight. I was nothing to him. I wander the world, digging my fangs into men's hearts to fill the void.

What makes the Kumiho different from other mythological nine-tailed foxes is the fox marble, or bead, containing all the Kumiho's wisdom and power. The marble is cream-colored and slithers in and out of the object of affection's mouth, consuming their life energy.

♥

I redownload Tinder and open Hinge. Running on the last bytes of data, lying against a mattress without sheets, I swipe right on a Chinese graduate student.

♥

There's a picture of Ahri, the nine-tailed fox from *League of Legends*, framed in the glass. She stares with chocolate eyes. I wonder if the orb in her grasp is my soul.

♥

The Chinese graduate student's lips hook onto mine the moment I get on his bed. The bead goes in and out of our mouths. His mouth reeks of pizza. His legs prickle against mine. He wants more.

"You're not going to ghost me, right?"

A four-hundred-dollar lamp, fast Wi-Fi, a thousand-dollar desk, and a car.

"No, of course not."

Sometimes, a man's heart isn't enough. I want the finer things, too.

♥

"You're an Ahri bot?"

"Back in the day."

"You're *that* player."

♥

Cute Physics Boy lets me touch his injuries. They're all on his left side. I trace my fingers on his left hand. The nerves don't work anymore. I'd like to saw off my left hand and sew it onto his with red string.

♥

I'm nicknamed Ahri by my fuckbuddy. Ahri and I are one and the same.

♥

"You're just like me. You run away when someone says they love you."

I am nothing like my mother. I'd rather have one of my nine tails burned to a crisp.

♥

"I'm sorry about everything. I'm sorry about my actions these last few months."

I love him. I love him. I love him. It burns when I text that a simple apology won't be enough anymore. I want to be treated like a decent human being. I hide underneath my blankets. I listen to the rain for a few minutes. I cry. Whose heart will I devour tonight?

♥

In *My Girlfriend Is a Kumiho*, it rains because the Kumiho cries. Her boyfriend betrays her.

♥

The first time I fall for a woman, we lie next to each other on the bed. I want to see her grow, to become an alluring human being who is loved and loves. I want to run a finger through her hair, kiss her, and tell her how much I aspire to be like her. I kiss her. She tells me kisses mean more than sex. I apologize.

♥

Swipe left. Skip on his heart. Swipe right. Eat his. Repeat.

♥

"I have four other friends with benefits," the fuckbuddy tells me.

I laugh. We lie under the stars. The waves dance. Staring at the stars on the beach is romance, but is it us? There is no desire to be in love with him. He talks about them. I listen. On the third date, he tells me about his other lovers. He's stuck with me until he inevitably catches feelings for one of them. I'll take his energy before it happens. It won't be me. I only have a heart cavity. I eat hearts.

♥

Screening process on Tinder: If they get baited by the line, drop them, unless they're worth the fuck. If they manage to entice you with conversation, move to texting. No dates until a week of texting. Repeat until you only think about the heart you're about to demolish.

♥

My half-sister texts me to be safe after I talk about dating apps. I laugh. I have five years of experience under my belt. I've used dating apps recklessly and carelessly. Not minding the consequences. Not concerned with safety. If someone harvests my organs, let it be known my liver is the most delicious.

♥

Ahri seduces men so she can eat their souls and steal their memories. I don't want their memories. I don't want to remember them.

♥

I want to text the engineer saying he's going to fuck all the feelings out of me tonight. I erase the message because the engineer is a banana. I'm not willing to pull back his peels to see if his fruit is as delicious as the cum I will swallow.

♥

I cry on the fuckbuddy. Not because I'm in love but because I'm too lonely to fuck. When he puts his dick in me, I yelp in pain. Why does it fucking hurt? There's sadness building in me. I want to read. I want his fingers to run through my hair. I want him to kiss my body as I sound out every single Vietnamese syllable. Instead, I pull out Hinge. Instead, there is a red outline on my ass.

♥

Hinge

> I should invest in more chokers.

She's so pretty.

...

I cannot screw this up.

...

> Or you could let me choke y

My fuckbuddy presses send, not me.

"Why would you press send?!"

♥

A new boy bought a sweatshirt in Ireland when he was visiting his ex. I put it on. There are still parts of her lingering on the sweatshirt. He makes coffee, I play music, and sunlight seeps through the window. I want parts of him.

♥

Relational Fate: People meant to be in your life will appear without any effort. It's the red strings of fate. When the new boy tells me about it, we both chuckle. How do you believe that fate will bring two people together without work? How do you maintain a relationship without work? In the dark, holding his hand and waiting for the light to turn green, we bubble in hilarity. I like red strings of fate as a concept, but I don't believe it like I don't believe in romance.

♥

I'd like to think of the boy I gave my heart to, the boy I'd give up immortality for, as Nymph. His world was tied to my version of Olympia. Fascinated by him, I followed him downtown. I weaved and bobbed through obstacles. I never knew about his problems. He had unresolved issues at the time. He pushed me onto the concrete, scraping my face.

♥

I want a rose tattoo on my back shoulder. I want more money. I want to eat their entrails. My lips will be a glistening red others envy. I want, I want, I want.

♥

My sugar daddy leaves an envelope on top of the rice cooker. I tear it apart. I hide the hundred-dollar bills inside my glasses case. No one needs to know my vision isn't perfect. No one needs to know this money is too perfect.

♥

Today, I persuaded a boy to buy me groceries, cook me lunch, and put together my IKEA desk. I've enticed him with my tails, pretty rose tails perfumed with lavender.

♥

My fuckbuddy picks at the scab on my arm while we lay in bed. His fingernails scratch at me. A part of me wants to pull my arm away. I want to pick at my own scabs. I don't want a foreign entity scratching my wounds. These white reminders are mine.

He mimics my anxious tick. By imitating, he's taking steps to own me. He scratches at my skin like it's his. It's as though he is entitled to my flesh by scratching the scab.

As his fingers graze then attach to my wounds, his actions remind me of my distractions. I don't have to think about the boy I love or the people I want to talk to. He pulls his hand away. I pick. I don't need to tell anyone about the men in my life. I don't think I should let the blood stream down my arm, but I pick anyway. I see red on my arm.

♥

One of the boys unfriends me on *Snapchat* when I tell him I can't hang out Thursday. I don't need his dick. I'll have a new liver to snack on next week.

♥

I want to hold your hand and eat your pussy.

♥

Cute Physics Boy is with his friends again. He notices me first. A part of me craves his attention. I want to entice him into giving up his left ribs so I can replace them with mine. I stare over his shoulder, my hands doing what they know best.

"Cancel Spotify?" I ask as I stare at his Google Calendar.

He tells me he needs a reminder to cancel his free trial, but I'm distracted by his ribs. I want those ribs.

♥

I miss you. I miss you. I miss you. I keep rewriting texts, hoping you'll come back. *I don't want anyone but you. Please, please, please. You might be the reason why I give up immortality. Let's replay our memories. Let's replay the moments I put off devouring someone else's entrails so I could be with you.* My tails are painted rose. My lips are succulent. What else can I do? What else do you want?

♥

> How was moving in? No more ants?
>
> Read at 9:47 AM

A liver a day keeps the mortality away.

tumblr

Anonymous

Anonymous asked:

You probably know who I am, but I remember you confessing to me, and I was and still am with my girlfriend. I really enjoyed our conversation and wish we had more chances to hang out before you left. Maybe if we were both single, things could have been different, but I blocked you to get over you as temptation. I wish we could have stayed in contact, but I made a dick move.

I wish I had communicated with you. I value you as a friend and hope you still want to be friends. I'm sorry about how I went about things, but it was a really difficult situation. Anyways, I hope we can still talk.

"Wow, you look hot."

Don't look at me like that.

♥

The second time I fall for a woman, I hold her hand. I sit next to her at a concert. We're awkward. I've never felt like this before. The uncomfortable silence, the fear of being rejected, not knowing what to do with my hands. We're both at a loss for words, but I ask if I can kiss her. My lips find a home on her cheek.

♥

"To be honest, I could do a long term with you."

"What?"

"But you've told me all your boy stories."

♥

September 17: 9:56 AM

> If you're down, let's go out sometime! :)

September 21: 11:23 PM

> Yeah, maybe you can show me why I shouldn't save myself.

Read at 11:35 PM

♥

The fuckbuddy, who is my first San Diego fuckbuddy, blocks me on social media after I express my emotions. He takes it as a personal attack and tells me, *look at all the things I've done to make you feel comfortable. Why aren't you comfortable?*

♥

I post on Instagram.

I cry realizing I truly believe I will never be loved without objectifying myself.

♥

I must return the fuckbuddy's sweatshirt. I don't want his parts anymore. I've taken my fill and more. I write thank you for being my prey, apologize that the friendship ended the way it did, and write his name on the note. I leave the note and the last part of him outside his door.

♥

tumblr

Hi anon,

It took me about 15 minutes to figure out who you were because you never gave me context. Also because I deserve more apologies from guys like you. Y'all really think you're special or something. You gave me fragments and pieces to deal with, and never told me you were with your girl. To me, it was implied y'all were talking not dating.

I never had romantic feelings for you. I never had feelings. I don't know why it was assumed I had them or when you thought I confessed to you. Three years ago, when you rejected me for coffee, it was enough for me to think, *nah, he rejected being friends, so we're not doing this.*

I only ever wanted to be friends. That was it. Which is why I said I just wanted to check-in on you. That's why I always replied to your posts. I was just caring about you. And apparently that was too much. We both know that there wouldn't have been extra opportunities to hang out because we didn't start talking again until a few months before I left. And that's the thing, you only value me checking-in on you. I don't think you valued the friendship. Y'know why? Because if you did, I wouldn't have been blocked. I don't remember you ever checking-in on me, I don't remember you ever starting a conversation. This was never a two-way street. And so, yeah, I'm not surprised you blocked me. I don't think you ever treated me with the same respect and value I gave to you.

♥

Gay panic. Gay panic everywhere. There are too many beautiful girls. Kumihos only care about boys. Only eat the boys' hearts. Leave girls alone. They are hurt by so many things already. I don't want to be another reason why they hurt.

♥

My first San Diego fuckbuddy unblocks me to tell me he's open to talking to me in person. I don't need his poisoned liver. I don't need his broken heart. His pride will give him a heart attack.

♥

My drunk-self convinces a boy to break up with his on-and-off again girl so I can be his one-night stand. The power.

♥

"You can leave now."

"Not again, I thought I left those bitches back in undergrad."

He wraps his arms around my wrist and pulls me back into his bed. I laugh.

♥

The Nymph blocks me on Tumblr.

♥

Sugar daddy catches feelings. He doesn't know about the tails, the liver I yearn for, or the blood staining my lips. I don't want his romance. He's not worth having humanity.

♥

Once upon a time, you saw my true form and ran away.

♥

I'd rather be the villain. Tails perfumed with lavender. Lips red from oozing digestive juices. Fingers tainted from breaking ribs.

♥

The day I learn I don't have to worry about a teaching assistantship, I tell the new boy, an almost crush, I'll take him out. We watch *Parasite* together. We buy groceries at Ralph's. We discuss our relationship.

"I don't want to put a label on this," he starts.

"I'm not ready for a relationship," I continue.

"We're like regulars," he ends.

I don't care. I still want his ribs. I want to tear open his chest with my teeth, bite down on the flesh as it reveals the sternum, the rib cage, and the beautiful lungs. I want to hammer away at the sternum, watch as the ribs crack little by little. Yank at the bones as the sound crackles in my ear. Let my tongue glide across the tissue, tonguing the softness, searching for the area with the most pleasure. Tangle my tongue into bone and tissue. When thoroughly aroused, I want to sink my teeth into the spongy tissue and suck the marrow.

♥

                                          ME

                        9:59 AM

> Yo, let me buy you a beer next week, yeah?

CUTE PHYSICS BOY

      3:47 PM

> For sure, been busy with school that I haven't grabbed a beer in weeks. Appreciate it though.

I want one of the seven false ribs on the left as an entree. His heart will be a nice dessert.

"You're head over heels for him."

I'm terrified I'll want to sacrifice my perfumed tails for the almost-crush.

My fuckbuddy was a disaster. He was also convenient. He would interrupt my stories every five minutes. Every time I tried to unzip his skin, he'd let me know his flesh was gamey, but when I cried over the mortal I wanted, he was my only friend. He was a twelve-minute walk.

He had other fuck buddies. It should have been easy. I didn't need to be with him. His existence was enough.

♥

You catch more hearts with honey than vinegar.

♥

Growing up, I always wanted to be a princess, the girl with the beautiful ball gown. I was the sleeping beauty who princes kissed. When I imagined romance, I imagined ballrooms with slow dancing. Growing up, I thought the prettiest dresses were blush pink. When I started my adventures in romance, I believed otherwise. I watched as princes stole treasures from my tower. I naively took their kisses as trust. Spending years in romance, I acquired too much wisdom and my rose-tinted naivety disappeared. They crushed my heart. In return, I massacred their hearts.

Instead of becoming a princess, the universe granted me immortality. I entranced boys into slow dances, I batted my eyelashes, and my French-manicured fingers brushed over their rib cages. I wanted to unlock their heart's chambers. I exchanged my priceless, pure heart for eternal beauty, a fox marble, and powers. The universe, understanding this poisonous wisdom and knowledge, gifted me with pretty rose tails. I inherited a conventional beauty. I looked gorgeous in carnation. Still, I enjoyed the simple pleasures of pretending. The waltzing, the flirting, the dew of romance.

For the rest of eternity, I will quench my thirst with men's cum, livers, and hearts. I'll suck their life energy with kisses. Once upon a time, I believed in some silly rendition of romance. Now, any time I question immortality, I apply blood from the mitral valve as my lips shimmer. They crushed my desire into a powder, so I sink my canines into their dicks.

Swipe. Swipe. Swipe. My tails swish with every motion.

THE VICTIMIZED LIVER

The modern Kumiho chooses suburbs over cities. The pavement there is a little clearer. The moon is a little brighter. No one believes in devoured hearts. No one expects the Kumiho, a mythical beast, to prey in dead neighborhoods. Suburbia is easier because digging into men's chests with iridescence and leaving desire on their necks doesn't allow for suspicion.

The modern Kumiho turns off the living room light while the moon is high, while the streetlights flicker, while unsuspecting picture-perfect families watch television and share their lives at the dining table. She leans in to kiss her victim and digs her claws into him.

In the darkness, she slides her finger down his chest. The static of each home screams louder than her fingers carving into organs. They tune out the victim's orchestrated misery. When she tiptoes into the kitchen, blood sprinkles the hallways. She cleans the liver with cold tap water, boils it for a few minutes, and cuts it into bite-sized morsels.

When the birds chirp on the telephone wires and the sun starts to wake up, she leaves the door unlocked. She walks home. Her fingers are perfect. Her sneakers are wiped clean of any blood or poison. The families wake up and don't notice what's beyond their walls. If they did, they might suspect the garbage man picked up decapitated heads, limbs, and organs.

But the modern Kumiho chooses suburbs where energy and life flatlines, where each house's routine is predictable. Others may ask why she'd sustain herself in a sleeping neighborhood. She shrugs her shoulders, applies another coat of blood onto her lips, and tells bystanders to worry about their Tinder account. Dead and empty men, who have nowhere else to be, simply breathe. She relieves them of their banality as the white noise from electrical poles gossips outside their homes.

♥

Every victim is calculated. There is a type. The patterns are the same. Every victim has been meticulously chosen.

♥

"Do you eat that pussy?!" I shout while the ~~almost~~ crush and I walk back to his apartment. Our hands are intertwined and I am not tempted to skin them. I spin to hear an answer from the taller friend. If I was with anyone else, I'd also run after him, but my fingers are tied to the crush.

The clouds amble across the sky. My veins are rearing with alcohol and anticipating the blueberry donut inside the brown paper bag. I feel at home while suburbia snores at eleven p.m. My body finds peace walking in the middle of a road, laughing a little too loud for his neighbors, and screaming about eating a woman out.

"Every time!" his friend shouts.

I screech, alarming all the ghouls who linger among the dimmed streetlights to join the festivities.

♥

The Nymph slides into my text messages. His pretty words are chocolate-dipped and I lick at them like they're a dick I'm teasing: without hesitation or desperation, but with recklessness and intrigue.

He apologizes for feeding into my insecurities and not reading my heart. I'm awake for the rest of the night. I stare at the flashing Christmas lights. I look back at the texts in the dark. He apologizes for treating me less-than-a-person. He thinks he can treat me with the respect and attention I deserve. I continue to lick the chocolate off his words. He's not any different from other boys who crawl back.

I strip the chocolate cover, which hides intention, reveal the flesh, and sink into the pretty pink words. I spit them out the window and give him one more chance. My fatal flaw is letting men who put my heart aside crawl back.

♥

The first time my crush comes over, I ask if he wants to take a shower.

This is the first time I invite a boy to shower with me after our first fuck. This is the first time I want to keep the affection and desire lingering in the water molecules. I kiss his back. I kiss his lips. I keep desire.

♥

I have Chlamydia again. I sit on the hairballed carpet where my tears pool. I did everything right. I had safe sex. I made sure everyone wore a condom. I did. I did. I did so well. I text the crush. Frustrated and crying, I pick up another prescription of antibiotics.

One of my other partners calls to ask me how I'm doing. My body heaves. *I can't do this. Not after someone makes out with me in the coffee shop and equates my intelligence to sex appeal. Not after a guy tells me to shut up when he picks up the phone. No, no, no more please. No more boys who belittle, demean, and treat me as a thing. I don't want to be played anymore.* He comes over to pick up the partner dosage.

"How many other guys did you give this to?"

I feel a jolt of insecurity.

"I'm curious."

Is it curiosity or a death wish?

"No, we're not doing this. This isn't funny. This is serious."

He isn't supportive.

"This is my coping mechanism."

He doesn't care.

"Sorry, dude."

I stop caring about him.

He texts me a few minutes later to tell me he's pissed and that I shouldn't be fooled by his smile. He tells me not to beat myself up too much. He tells me he's always there to support his friends.

I don't tell him I don't care. He doesn't get it. This isn't support. This is some twisted game. I'm not a novice. I gambled my vulnerability and was picked apart by another man's ignorance.

♥

I wish I could suck all the life energy out of men with one kiss. I chip away at it instead. The fox marble can only steal a little wisdom at a time.

♥

My guardian angel accidentally falls to earth and I see his wings for the first time. Molted and grey, he is not the same angel I crushed on a year ago. I take a feather and as he circumcircles in speech, I pin the feather to my ankle. He stares. I grin.

"I hope you're not here to judge my lifestyle. That would be kind of rude of you. I know you're kinder than that." Off-put by his words. It feels like there's a different intention behind them. I unpin the feather from my ankle and stab it to the back of my thigh. I feel the blood trickle from the pin. I push a coy smile onto my face.

"I don't have time to judge you, but you have said concerning things I can't put aside, considering our history."

I stare at the fallen angel. I smell the iron. The honesty is gone, replaced by fear. I let the wound pus.

♥

Why did I have to watch you choose someone else over me? Why was I never your first choice?

♥

My first San Diego fuckbuddy slides into my DMs and apologizes for treating me the way he did. He tells me I was right, and that I was just trying to be a good friend. He wants to be friends again. I pause to apply a chocolate balm to my lips, a ritual to leave true intention underneath a dark dip. I want him to keep talking. If he stops, I don't get to indulge in the humiliation.

I want him to apologize for always trying to have the last word. It rips all the courage from my jaw when he does. He doesn't deserve the last word. He continues to speak like he's greater than me.

I try to reply as kindly as I can, but I taste the balm on my lips, a constant reminder to leave an authenticity and honesty behind a rich delight, to never let a man taste my lips. I accidentally bite my tongue, tasting iron. He says,

> It looks like you're enjoying your time in San Diego. Well done.

I lick my lips again. What is the true meaning behind those words? I'm unsatisfied with that message and unsatisfied that he continues to say useless things. The blood stains my teeth. He tries to steal my limelight.

> Never talk to me. Keep your word and let that be the last message. This is a pattern of behavior that no one should have to go through just because you get defensive.

He blocks me afterwards. I cannot taste the flavor that covers his venom. I only feel the poison corroding my teeth enamel.

♥

If I can't be the love interest, I'll be the mistress, equipped with heels, a corset top, and skintight jeans. If I can't have you then I can paint myself to be something you always desired.

♥

When trying to find a Kumiho, go to Tinder. She sits on her bed in some god-awful apartment with three other roommates who also are talking about love. She swipes on potential prey. While ambulances wail in the background and there is occasional creaking from the floorboards upstairs, she vets men on her list and questions their heart's desires.

The warning signs that she has chosen her victim include her starting conversation, being too invested too early in the relationship, and not giving you her number even though Tinder is laggy.

On the first date, she meets you at your apartment. It might be in broad daylight or in the middle of the night, and when you inch closer to her, she immediately edges closer to you, too. She makes you feel comfortable, so you believe slamming her head onto your dick is a good thing.

Beware of bite marks. Her blowjobs include teeth: pointy canine teeth. You're probably wondering how a girl is that bad at blowjobs. Her aftercare includes kissing your collarbones, nose, and lips. She apologizes for her skills and hopes you invite her again to work on them. When you let her go from your apartment building, she will run a finger through her hair, flash a goodbye, and hop into the Uber you bought her.

She's looking at her collection of recipes as the driver asks how her night's going. She selects a particular method to eat your liver. The city blurs as the drive continues. She returns to her mattress, replies to other potential victims, and lies about how great the date was. The city is awake. Her bedroom window reminds her about reality, wilting roses, and shitty playgrounds.

I confess my feelings to the crush. I tell him this isn't a discussion about our relationship. I still don't want mortality. He kisses me. We go back to the bedroom. I stumble on my words. I gift him all the poems and prose I wrote about him, all bound by red string.

"Some evaluations will be based on the fact that you're a young female teacher."

I bite my tongue. When I go home, I lick the pretty off. I've spent so many years perfecting how to be desirable. I won't be fazed by fourteen boys telling me I'm pretty. At least I'm in control.

♥

The difference between sending nudes and being objectified is simple. I'm in control when I send my nudes. I control when and how I should be desired. I love being the center of attention. I love watching people jack-off to me. Still, there are instances when I like to be objectified. Is that so wrong?

♥

"Interesting."

I can reply to my sugar daddy or I can leave him on read. He can continue to assume I moved on. It's been three months since I've heard from him. I'm too occupied with my own frivolous matters to care about him. The feather digs into my thigh. I reply.

♥

I try to muster-up the rest of my energy to talk to this guy. Recovering from drinks and wingwomaning my friend, I think everything will be okay, pleasant, kind. Except, he kisses me without consent. He kisses me without consent and his slimy lips are still stuck on my face.

He kisses me he kisses me he kisses me.

He tells me I was so shy but so flirty when he talked to me. That doesn't make sense because we only talked about academia.

He kisses me he kisses me HE KISSES ME.

NO. NO. NO. WHY WHY WHY?

♥

I fall asleep on my crush's couch. My anxiety evaporates. He pats the spot next to him. I lay there. My stress dissipates as he wraps his arms around me. My eyes get heavy. I take a nap to *Bob's Burgers*. This is the safest I've felt in the last two weeks.

♥

When I collect the essay drafts, two boys from the second section impatiently wait for me in the back row. I am ready in my battle gear. I've been prepared since I woke up for freshmen who don't know how to write essays. A white crop top with straps on it, white, high-waisted skinny jeans, and white sneakers. I change personas as quickly as I can. I stare at the boys who tell me their papers are full drafts. I nod, not really caring, knowing people's words about being a young female teacher.

I catch the glimpse of desire in one of them. Grinning, I collect their drafts. I'm a young woman of color with a cheerful disposition. When I go home, I'll suck the pretty out, unpin the feather, and clean my wounds.

♥

We talked about academia. We talked about graduate school. He mansplained TA'ships. We talked about the lack of diversity in grad school. I realize my intelligence was objectified. The worst part is knowing a man can twist an admirable trait and make it an object for the male gaze.

I want to go home. Please, please, let me go home.

♥

The afternoon after some boy non-consensually makes out with me, my guardian angel and I get into a bad discussion. I can feel the feather dig into my thigh. I call him. He picks up.

"Don't say a damn word," he says.

I don't.

"...I don't even know what to say..."

I want to yell, *I called you with something to say!* I want to say, *Why would you tell me to shut up if you have nothing to say? You aren't a bad guy. You aren't toxic. I can accept responsibility for not trusting you and being hostile and defensive, but why, why, do you get to say something first even though I called you? Why do I have to be quiet again? I don't hate you, but I am disappointed and hurt because you never explicitly stated you threw me away for two other girls. Two other girls, not just one. Why? I know I've done wrong, but you want to be an angel. I never wanted you to be an angel. I never wanted to be the human you looked over. I just wanted to be yours.*

I don't say anything until he's finished. It doesn't matter what he says. What matters is he wanted to hurt me.

♥

The problem with being a Kumiho is you let it happen. You let boys touch your tails, force themselves onto you because you want them to like you. You need that attention to stay immortal, and wake up the next morning still sticky from yesterday's undesirable touches.

♥

There are nights when the crush and I do not cuddle to sleep. On those nights, our backs are facing each other. On those nights, there is a terror growing behind my ribcage. It is a Kumiho instinct. It is a reminder to cut the strings.

♥

"The men who leave me always come crawling back."

"Didn't you like them at some point?"

The Kumiho sips her honey and rose latte. Her fingers dig into her skin. Outside the coffee shop, she watches as a couple passes. A garbage truck lifts the gray trash can. When they walked together earlier, someone stared at her trench coat then down at her brown Lita boots. Before that, someone whispered *beautiful* in her ear, and she replied robotically with *thanks*. She wonders if body parts are being thrown away.

"Yeah."

Her roommate discusses how confused she is about mixed signals from men, definitions of a relationship, and how he leaves her on read. The Kumiho replies to her sugar daddy. The world outside the coffee shop is gray. It rains and the cars zoom by.

"You don't let someone back in just because you liked them before. You shouldn't just be someone's *choice* when they were your *priority*," she states after her roommate discusses more about her current crush. The Kumiho stares back at a decapitated head, reminds herself to drill a hole into her next victim's skull and pour hydrochloric acid inside.

♥

Google Docs

Hi guardian angel,

You've gained access to this Google Doc. Please don't comment on this document that is mine. Please don't silence me. Please, please, please, don't. This is one of the few places I have a voice, one that I like, one that I accept, one that I would edit again and again in hopes that it doesn't make you look like a bad guy.

You aren't a bad guy to me. I'm the bad guy to me. Did you read between the lines? Did you read the implications? Or did you read this like you were still the toxic one, that there was a thorn in my side because of you?

I'll be the bad guy. I don't believe in altruism. I believe every single human being has an agenda, a selfish desire, and I don't believe for a second that you are taking yourself out of my life because it's good for me. Let's be real, it's good for you. The way I talked to you was dismissive, aggressive, and defensive. It was stand-offish. It was like the Kumiho I rewrite myself to be.

Why? It would be so fucking cliché if it was to drive you away, but I'm not a sap. It's because I wasn't ready for you to be in my life. I was still hurt from you teetering and tottering on girls in your life. I was the jilted lover, and you can bet your holy Angel ass I resented you for it. You broke my trust but did nothing to deserve it in the first place. So yeah, I was the villain. I was the thorn, and you don't get a fucking say in any of this.

You don't get a say in this narrative. You don't get to edit this and tell me how I wrote you wrong. You already had your diatribe in our messages. You returned to my life and pretended nothing happened. I wanted you to acknowledge my pain. I resent you.

You aren't this story's bad guy. I am. I'm sorry I was a villain to you, but you still don't get a say.

I hope the next girl you decide to be friends with knows about me. I hope she takes your side because I have been a wonderful villain in too many people's narratives. It's easy to blame me. It's easy to vilify me. Do it. Go for it.

♥

I have a habit of starting sexual relationships with tolerable men. I think cute guys are rare despite their terrible behavior. I block the "nice guy partner" so I can start changing my habits.

I'm terrified my crush is using me. I like someone enough to give up immortality. I remind myself to carve out his heart, saw at the skin, until my hands can dig into the cavity, feel the heart pulsating in my grip, feel the warmth emulating with each beat, and yank the precious thing out without hesitation. Afterwards, I remind myself to cut the heart into bite-sized pieces and let the ventricles' juices run raw before chewing away.

Sometimes, I still desire to be a princess. Sometimes, I want my paperweight heart back.

The nice guy apologizes to me by giving me a succulent. My roommates ask if I want it inside. I tell them no. I'm not taking care of a plant I never wanted. The next day, someone rips all its leaves off.

♥

My crush has another woman.

If I had a heart, he crushed it into crumbs. I thought he would be different. I thought if I revealed my tails to him, he wouldn't pull.

In the night, between smoking joints and smudging graphite on my callouses, tears stream down my face. The sprinklers sing outside. The static in my room deafens. I wish it would rain.

THE PAPERWEIGHT HEART

Does having a lover make mortality worthwhile?

I haven't found prey worth staying for. I haven't found the right fingers to interweave mine with the way stories tell me they should.

Are your breaths in sync? Do you feel warm with them? Tell me, were they worth mortality? Is love ever worth mortality?

♥

Instagram

12:51 PM

> Your Tinder advertised blowjobs.

> It was the biggest thing on your profile. Just fucking fuck off.

♥

If knowledge is power, every man I've kissed has left their knowledge inside my marble. Every time I swallow it, the bitter taste leeches onto my teeth.

♥

When my crush says he can only offer me a friendship because he has another woman, I try to find words to type to him. Instead, I dial his number. I only understand hurt as undigested hearts mutated in my stomach, crawling around, feeding off my pain and destroying my body.

♥

My sugar daddy has some clear rules about his dating life. He dates casually for four months, and then asks if his partners want more. If they don't, he continues dating other women.

I can't comprehend. Four months is too little time for me to want the ribs, the lungs, or the lips. I want to learn about my victims' bodies before I devour them.

♥

I visit Olympia after I'm heartbroken. The first full day there, I wear a blazer, bra, and laced up leggings. I stare down every driver who looks me over. I run a hand through my hair, letting the scent of lavender travel through the air.

♥

When I tell someone my crush is canceled, she tells me to write about it. I stare at the word processor. For once, I don't want to write about him.

♥

Another boy gifts me a succulent because he thinks I'm a plant mom. I stare at the ceramic corgi pot, glance at the blackened leaves, and wonder how wrong people are about me. I never consented to caring for this plant. I only care for roses. I love pricking my fingers on thorns.

♥

Instagram

> you have any other potential lovers you're talking to tho?

Why would I want to be with anyone right now when the crush I want so much more never told me his secrets? He didn't tell me I would never get close. I could have eaten more hearts if he had. Why would I ever, ever want to be with someone else right now?

ME

> Nope! Not really looking for love.

EX-OLYMPIA-FUCKBUDDY

> Not love, just a little fun. Y'know sparking some romance?
> Some flirting?

It's lonely only having sexual partners over, but I don't want to look at anyone like that.

Alone at night, I don't want romance. I don't want another body on me. I want warmth, tenderness, care, and love. Why can't I have that? Why does it always have to be romance? Can't it be love in the form of friendship and conversations?

> I don't want to be flirty. I don't want someone's company like that. I just want to be with my friends.

♥

How many times will I beg for someone to care about me?

♥

One of my partners is known for always supporting his friends by sending encouragement, going out of his way to drive somewhere or buy them food. During one of my worst depressive episodes, I send him a message telling him I haven't eaten all day. He replies he hopes I get some snacks.

The next day, he offers to come down and spend time with me or to accompany me doing work on camera. What I really want are snacks. What I want is someone to make me food because I can't take care of myself when my heart is being devoured from the inside out.

♥

Does your human make love worth living for?

I've been dying because of love.

♥

Please tell me my worth won't always equal my physical appearance. Please tell me when my looks disappear, I'll still mean something.

♥

Day 20 of Quarantine

I delete your number. I'm sorry. I can't rely on you anymore.

♥

"I dare you."

I'm a slut for dares, pretty artsy boys with good style, and useless playthings that make me feel attractive for a little bit. He dares me in my DMs and I remember why I eat hearts and livers.

♥

I delete my Snapchat, again.

♥

MESSAGES SENT TO MY DMS

> I wish I was single again because of you.

> What's it like to turn men on without trying?

> dtf tho tryna get over an ex.

> Why don't you send me nudes on your birthday? Y'know because your birthday.

♥

July 31st, 2020

I wish I still had your number. For once, I wish I could be as loved as you, guardian angel, without worrying I'd be thrown aside.

♥

"Are you okay?"

The Nymph is the first one to check-in with me after my crush almost steals my immortality. He is the first one, and between tears, between revealing all history, he listens. It doesn't rain this time but the sprinklers remind me of it. In the dimness of eleven p.m. phone calls, I cry and cry, forgetting that months before, I was crying over *him* and how I lost him. Now, the Nymph is someone who will always be a part of my life. Now, he loves me.

♥

My half-sister texts me, "Be safe."

I reply, "I have been."

♥

My ~~crush~~ ~~ex-crush~~ ghost posts on his Instagram story for the first time in my existence, and I stare at the slomo, listen to the Trey Songz song, and watch the blurry figure holding the sparkler. I feel my heart break. Is it who I think it is? I view it again. Fifteen seconds of a boy who I gave my heart to for ten months, who I let see my tails, with another girl on Instagram. I thought my lungs could handle it. I thought I could hype. I thought I thought I was over it.

I send him a message.

> Oh my gosh this is so damn cute. AND IS THIS WHO I THINK IT IS?

I send another to my friends.

> Ghost posted-up his girl and my heart hurts. It'll be okay. It'll be okay. It hurts, but it'll be okay.

I thought I was invincible to heartbreak. I wanted to be the Kumiho who didn't care. Why can't I be? Why do I have to hurt again? What is this cycle? How do I get out of it?

♥

Heartbreak hours are over. It's a heartbreak year.

♥

In a Snapchat story, I proudly present a bottle of wine, guzzling it down and gyrating my hips to the song because fuck this shit. My friend replies, "YES SIS THROW IT BACK." And all I can think about is how her encouragement is much more important than other replies.

A boy I'd like to eat the heart of sends the side-eye emoji and I reply,

> I'm trying to work on my booty.

> I'm definitely not mad at that.

> I might need your help, babe.

The ghost also replies. I don't open the message.

♥

"You are so fucking cute, but I can't flirt with you right now because of my mental health. Like fuck, I wish."

♥

My back hurts. I erased my back the first night I learned the relationship would change. It burns. In the dim light, I took a pencil to my back and erased and erased until all the sadness went away, until the pain replaced the heartbreak. The first night, I hurt so much. The anxiety wouldn't stop playing. The next morning, I turned around to see two large friction burns facing each other. They were lovers. The tension burned me.

♥

"I think it hurt more because I was more involved with her."

The ghost doesn't know how much I destroyed myself.

♥

One of these things is going to happen.

I won't find a partner I want to be in a committed, monogamous relationship with and I will end up planning a wedding to marry myself because I really like wedding dresses and want to have three different outfit changes

OR

I will find a partner and everyone knows they will be the end-all for me. I will host a party for my first-ever committed relationship, not inviting my partner because I need to celebrate this monumental moment in my life on my own.

♥

"BITCH, WHERE HAVE YOU BEEN ALL MY LIFE?"

It turns out I forgot who I was while I was invested in the ghost. I was reserved, quiet, and polite. I will never have the same conversations with him as I did when we were talking on Tinder. I will never be able to puncture his skin. It turns out I forgot how to be alive and confident, but then a boy reminded me of how beautiful mortality can be.

♥

> Hey, I'm so sorry, you're fine as fuck, but I can't continue fucking you for my mental health.

♥

The sun sets as I cross the street. Watching a cloud of bikers appear, I gravitate towards them. As I swing my bag of groceries, my eyes glance at the girl who sits on top of a motorcycle, her pink helmet stands out as darker colors paint the sky. I cross the street. I glance at the bikers.

Someone looks at me again and again. I try to stop the tails that perfume my mind, but I eye him. I eye the glasses. I look at his black motorcycle jacket. I listen to the whirs. A grimmer sky. A game of looking. A few seconds on him, a few seconds off, trying to catch him looking. It's my favorite game to play with strangers. Is he looking or am I looking?

The leader's bike roars even though the light is still red. One of the bikers follows. I wave. He looks back, I wave again. Someone else watches our game. The spectator looks at me. I put away the tails but keep the coyness.

"Would you have gotten on the bike?" the spectator asks, as we wait for the light to turn green.

My pockets are heavy with the remnants of my last apartment: a forgotten hairbrush and deodorant. A bag of groceries in one arm and my phone back in my new studio. I don't have to think, though.

"Yeah, I love a good adventure."

"If you had asked, I'm sure you could have gone."

I wonder if the stranger is right, that the biker would have let me on. But then would I have turned back into a Kumiho?

We go our separate ways. I think about how no one would have known if I disappeared on the bike if things went awry. I think about my recklessness and rashness in flirting, in romance, and in life, and sling my groceries back and forth.

♥

> Hey, sorry for ghosting you, but I don't think this relationship is sustainable and I'm not in a place to be casually fucking for a while. I'm sorry, you're still one of my best sexual partners, and you'll haunt me, but I'm not dragging you into my problems.

♥

The first night at my new apartment, I look out the window. I see the barely visible lights and my mind replays a memory. I thought the trauma was behind me. Instead, I remember being naked on a partner's floor, crying against the drawers, taking-in everything he lashed against me.

"You're terrible at fucking."

"Did you give me an STD?"

I remember being nude. Remember being drunk. Remember the strobe lights blinking in the background. Remember Seattle taunting me. Glancing out the window, my body shivers, terrified this cycle will continue.

♥

SUGAR DADDY

> I get it, and I hope things are looking up. 2020 has been rough for everyone. I know we haven't hooked up much, but I wish you the best, person to person. And when you're in a good place, you know where I'm at.

♥

I hop onto the SoCal Asian Discord, slightly intoxicated. I scream at the world in the voice chat. I have discussions about health inspections and flirt with people who have a beating heart. I haven't met new people in so long that I forgot what it means to introduce myself to someone new. I have forgotten what it means to be a human and not a Kumiho.

♥

The ghost sends me a meme on Instagram. I respond twelve hours later. I type, retype, rewrite, edit.

> Hey, are you free next week, I'm flying back to Olympia for my birthday and I'd like to hang out before I leave.

He tells me he'd have to ask his partner. The switch flicks. If I can't talk to him in person about everything, I'll have to use this digital space to tell him, to reveal all the tails, and to push his fingers away.

I tell him everything. I tell him how it hurt to read that my pain was less than his partner's. I send photos of self-harm. I listen to his responses. I tell him because no woman deserves the pain he put us through. No woman needs to be reminded of their insecurities because he chose to be dishonest. No woman deserves pain because of her partner's lousy decisions.

♥

I will always live my life as a Kumiho. I live as a myth to my partners' women. I live as an evil fantasy.

♥

I'm biased. I know my lungs will beg for my ghost's breaths. I know my mind slithers with thoughts of his heart and not mortality. What I want is not what I need. What I need is him to change. What I need is his other partner to be safe and grow. What I need is not what I want. I make a choice. It burns my heart.

♥

I know the Kumiho is a foil. I know it's a character. I know I'm not a Kumiho who eats hearts and livers. But when I learn as a young skinny Asian woman that I push-at people's insecurities, I am conscientious of the mythological role I play. I'm a fantasy to people, not only men but women. I exist as a villain. I exist as an object of romanticized desire for men and as a symbol of distrust for women.

♥

Putting on my mask, I walk around the mall, hoping to find someone who's going to this boba hangout. For the first time in several months, I am staying calm among friends. I joke around with the group underneath the orange sun and blue skies, waiting for a plum lemon green tea. No boba, no tapioca, just green tea with easy-going conversation.

♥

I love him. I will always love him. The ghost has always been someone special in my life. I let him shower with me the first time he came over. I invited him to cultural festivals. I almost missed classes because of him. I brewed him coffee. I let him meet my friends.

I knew he was different. Every moment I spent with him has been with warmth and kindness, as if I might lose him after every encounter. I didn't want to repeat my past mistakes. I didn't want to pretend I'd always have him. I didn't want to pretend because he was a ghost. He materialized when he wanted to. He haunted my thoughts and dreams without meaning to. When I was close to holding his hand, he'd vanish. When I tried to follow him into the dark, he wouldn't let me.

I kept searching for a phantom. I kept trying to wade into the darkness. I kept wanting to know more. I tiptoed in hallways to see if I could catch him off-guard. One moment, I crept-up behind him and saw him. I blinked. He was gone.

I spent too many nights with him. I spent too much time with his friends and him. I spent too much daylight with my body on his bed. I always played with his hair on the couch. Sometimes, he fell asleep while we were watching shows. In my wildest imagination, we were never together.

In my dreams, I take him to a dance battle in Seattle and show him who I am. In my dreams, we play video games and I hold his hand. I don't eat his heart. I don't eat his liver. In my dreams, I love him and he loves me.

He loves me by appreciating my efforts. He loves me by asking to see the writing revealing my vulnerabilities. He loves me by sending good morning texts. He loves me as I force him to try on clothes from his closet. He loves me by swallowing the ambrosia I brew for him.

He loves me because he watches me dance. He wants to see me as a human.

♥

My new friends invite me to a Google Meet. I happily accept. What transpires is an intense game of *Quiplash*, where Blind Jesus only writes WAP as an answer and a good way to know you are in a haunted house is when Blind Jesus talk-sings WAP. I learn my friends are switches in bed, and the dirtiest Truth-or-Drink answer is that one of our friends can make guys cum in under a minute. From eight to eleven-thirty, I learn about my acquaintances. I laugh and scream as loud as I can.

♥

On the last night in my first apartment, I tear apart the desk a boy put together for me. Red wine dribbles down my lips. I loosen the screws. The storage part falls off. The particleboard doesn't. I moan as I twist the screws. When I give up, I rip the boards apart with my bare hands. I watch the flimsy, dark wood pull apart with its screws intact. I jerk the legs. I twist limb from limb. I carry piece by piece to the dumpster. I throw everything into the air. The moon serenades me as I dispose of the last parts of me.

THE WIRED STERNUM

"In the case of Vietnamese, it is possible to clarify, to quantify the meaning of love through specific words: to love by taste (thích), to love without being in love (thương); to love passionately (yêu); to love ecstatically (mê); to love blindly (mù quáng); to love gratefully (tình nghĩa)."
– Ru, Kim Thuy

I came across this passage and burned it into my memory. For the first time, I found definitions of love that resonated with me. I didn't want to forget it. Quantifying love still seems incomplete, but this is closer to the meanings I yearn for. There is no finite version of love, but this is close.

♥

I visit my new studio right before I move in. As I walk back to my apartment, I stare at the palm trees that aren't native to Washington. The greens, browns, and oranges. I remind myself as the dirt kisses my soles that this is a new chapter. This will change. The sun is too high in the sky. The blue paints my ceiling. Things will be different.

♥

Someone asked what makes the Kumiho fundamentally different from other mythological creatures. She told me to make her my own. My Kumiho takes vulnerability and throws it toward hedonism. My Kumiho is one-liners, punches, and every single little thing I have ever been terrified of. She is me and reminds me that I don't want mortality. I want to pretend my insecurities, fears, and desires of being loved do not exist.

Still, there's nothing fundamentally different about her. I'm just holding onto a myth. I'm holding onto this red string hoping there is someone out there who won't pull on my tails and will make me believe it isn't worth the try. I'm trying to figure out if letting people in, letting them see how much I want them, keeps them scared or keeps them here.

Being this Kumiho is another way of pretending, wanting, and needing. Even if I want to eat everyone's hearts, I don't think I could ever hold someone's heart, feel their warmth against my palms, and think they want me to carry this burden. I only know how to gorge on hearts, not adore them.

♥

In trying to understand *League of Legends*, I ask my friend to explain Ahri's role. As my friend drives us through Los Angeles traffic to San Diego, he explains that Ahri is a mid-laner, which makes even more sense.

Each character or champion can take one of three different pathways, known as lanes: bottom (bot), middle (mid), and top, to get to the enemy's base. There is also a jungle that contains a gnarly monster inside the middle lane. The jungle is optional, but if a champion can defeat the monster, their team gains more experience points and, potentially, has an advantage over their opponents.

Mid is for those who can do well alone. It's for jungle. Ahri's abilities are meant for mid. My favorite *League* character is one who can jungle. Not only is she pretty, but she can handle her business by herself.

♥

I'd rather be alone and lonely. Not because I don't think I deserve a partner, but because I'd be stuck with myself. Knowing my insides is more important than eating someone else's. I don't need to suck someone's life energy with my fox marble to acquire that knowledge.

♥

My grandfather grew roses for my grandmother to show her his love. When he died, my father took over the rose garden. I think the pinks, whites, blues, and oranges are now a symbol of my father's love for my mother.

♥

My mother once told me I was like her because I run away from love. When I fell for my crush, I did not run. I was terrified he'd take my tail and I'd end up becoming mortal, but I did not run away. My feet, still safe in their Jeffrey Campbell Lita boots, stayed in place, then turned toward him. Step by step, my tails quivered because what if he was like the others? What if he wanted to take my tails for his own legacy?

With each step, I got closer to sacrificing my immortality for love. Despite my fears, despite my constant internal arguments, I moved towards him. I revealed I was a Kumiho and he stayed, though his hands wanted to pull. He wanted the lavender-perfumed rose tails. I had to hide my form again.

I didn't want to. I didn't want to think about his heart as my meal for immortality. It felt wrong to stare at his chest and think about his heart, or to be on my knees inhaling his third arm, always thinking about how his liver should be cooked. But when his hands went for my tails, I erased every desire for love.

I wanted to hold his heart, but he showed me I had to eat it.

I'm sorry, mother.

♥

I play with the fox marble with my tongue. It rolls around as I think about what I desire and yearn for.

I haven't given up on love yet. I don't think I can.

♥

Instagram

ME

> I know I can't blame you or anything, and that it isn't your fault. But holy shit, I've been dreaming about you for a week. Will you please stop haunting my dreams?

GHOST

> I'm sorry :(

♥

My first love visits me at my childhood home. It has been four years since I've last seen him. He greets my family and receives food. We sit at the table, talking, catching up on life and what we've been doing since 2016. It's like the old times except without the eighteen-year-old naivety, second guessing, and sadness. We talk until I force him to leave. As I walk him to his car, we hug like young adults.

We stare at each other like when we were younger, still intensely confused about love. I think there might be a flame. The night sky watches, and I know some sparks flicker, but what we had deserves to burn out.

♥

I sip my matcha as I fight an uphill battle. Months ago, I walked the same path with my pasts. They were both my different loves. Staring at the ground, watching as the sidewalk turns into road and grass, I wonder if this is how my life will be. I retrace the ghastly footsteps, but as I approach the streetlight, a different 90s song sings into my ears.

My feet dance on their own accord as the red light watches. The cars wait. I smile sipping on grassy matcha. My feet scrape ghosts' goo. The cars continue to move, and as I trek, I find the ocean, the cliffs, and the bench. I breathe in water. I breathe in the music. I swat the ghoulish goo from my feet. I will continue my journey without the ghosts. I follow my feet, which dance to their own tune.

♥

When my brothers pick me up from the airport, I find myself sitting in the back, enjoying the banter, laughter, and gossip they have for me. There's a heatwave in California, but in Washington, the sky is blue, the world isn't on fire, and I think about how the oranges, reds, and pinks appear in the rose garden back home, how cool the air feels against my skin in Olympia, and traffic jams feel more like love than people do.

♥

I have to put aside the Kumiho for a little bit. Not because I want to, but because I cannot pretend to be a heartbreaker. At stoplights, at bus stops, I put her aside. I dance, stare at the man who peers at me through his car window. I twist my legs, I work on my arm movements, and I glance at the man who looks at me like meat. A part of me wants to eat him, but I put the Kumiho away. The sun is high, the bus isn't here, but when the light changes, we nod. He zooms by and I continue to practice.

♥

"You live your life with such gusto and charm that it'll be a problem to form well-founded romantic relationships."

The Nymph read me.

♥

Months later, the ghost ghosts me. After a year and some change, he blocks me on all social media. I don't need a reason. The conclusion isn't important. We both fucked up, but I don't cry this time.

As I continue to reread old essay drafts, I glaze over the words. I will always love him. I wonder if there are dreams I can stitch together so those feelings of love stay sacred. Those moments, like my fox marble, are precious. And while my current view of him will always be tainted, the ghost will always be nostalgic and colored periwinkle.

His last text wished me good luck. I laugh. I don't need luck. I look outside and think it will be a good morning to dance. For the next hour, I dance to Vietnamese songs on the rooftop garage. I watch as the sky gradients into a new dawn.

♥

I go to my old coffee shop. I hear about how my coworkers miss me and how I pop up in conversation like an old memory. I smile on the inside, proud of my joy, proud of the baristas I've missed as well. I spark joy within. I pick up my cortado, pay, and sit outside the coffeehouse. I stare at the tiny downtown Olympia streets, smiling at the gray clouds.

♥

My ex-sugar daddy texts that he passed by my old spot. I smile and reply, enjoying the small talk. On top of the parking garage, I take a break from dancing to reply to his messages and learn that he has a girlfriend now. An almost happily ever after for the both of us. Him with someone. Me with myself.

♥

The Kumiho is different from other mythological, nine-tailed foxes in that it has an evil connotation.

While other nine-tailed foxes are tricksters, a Kumiho seduces men. A woman confident in her attractiveness is evil. A woman who claims her femininity and eats body parts is deemed the villain. Let me be that.

♥

On my birthday, my half-sister visits for dinner. Some parts are a blur and other parts are enjoyable. I remember telling her how my father roasted my dating life. I recall the phone call and how he thought I was always online dating.

"I'll let you go find more candidates," he'd told me.

I tell my half-sister this at the table while eating fried wontons. I turn to my dad, who sits in the twirly computer chair. I say I'm taking a break from online dating.

"Oh, so yesterday?"

I screech and feel the warmth rise to my head. I translate the words through his false teeth and smile so big. My father is the only man I can love daily.

♥

"Do you ever feel lonely?"

Maybe a year ago, maybe when I was looking for validation, love, and kindness. Maybe when I feared San Diego and I wanted a distraction from my anxiety. I objectified myself because I thought the only thing I could do well was date. Finding "friends" on dating apps was a distraction from my problems, my loneliness. I wanted to be loved.

♥

The ghost appears in my dream. Superstition says if someone appears in your dreams it means they're thinking of you. I'd like to think I'm haunting his dreams. He doesn't haunt my thoughts anymore.

♥

My last night in my old apartment, I stare at the black IKEA chair I put together last August. My lips rim the Merlot from nights ago. I remember sending the ghost a message about how proud I was to put something together by myself. I remember telling my ex-fuckbuddy about it, too. Almost a year later, I don't talk to my ex-fuckbuddy, and the ghost is a vanishing act. I grip the chair and take it to the dumpster.

The rusted, black dumpster is intimidating. A large chunk of metal looks down upon me in the dark. I think about jumping onto the edges of the lid to fling the chair into it. I think about being on my tippy toes and placing it gently into the mess. I stare at my slightly toned arms. I throw the black chair up and over. I watch as it flies, forgetting the faint tipsiness in my breath, and then gaze as it falls on top of mattresses.

♥

When my birthday party is over, a few of my friends sit inside the house, which is filled with my childhood photos, family, roses, plants, and Legos. As I look around, hearing the worst and best ho stories, I feel my eyes water. I wanted this. I needed this. The only time I'd sacrifice immortality would be if it was for these moments. I need *this* kind of love.

Halfway through one of my friend's ho stories, my eyes light up. My bare foot touches her bare foot. Among the group of five, I scream, "We're coochie cousins!" We downward spiral, analyzing that dick, different sex positions, and locations where we had sex.

♥

The last time I see the ghost in person, I realize all my feelings are gone. There is no more romance, no butterflies, no pink tint. I smile thinking about it. Love changes its forms and I am no longer romanced by him.

♥

With my eyes heavy and my body fatigued, the Nymph and I sit in a parking lot. We talk. The dim blue light watches over us as we exchange stories and discuss death, mortality, love, and fears. On the night of my birthday, we spend time together, just like two years ago. Except this time, there are no sparks. There are no dashes of romance, and I do not fall onto the concrete. There are no tricks to play. Now, he loves me.

♥

I cut my red strings of fate. I will not allow relational fate to determine who I love. I will use these frayed ends to murder. Tie these men up and choke them out.

♥

The sun sets as my best friends and I drive out of Oregon and into Washington. We sing along to English songs, bop to Vietnamese pop, and discuss our lives adjacent to growth. I don't know when the next time I'll see evergreen trees illuminated by the setting sun is, but I know as I stare at the freeway and empty spaces between cars that no amount of time and distance will unravel our history.

♥

Underneath the living room lights where my friends are, I remember how to love, remember how to feel, and remember who I'm willing to sacrifice immortality for.

♥

I wonder why loneliness is the Kumiho's punishment. As I trek up the hill to Sunset Cliffs, I wonder if being alone is a sadness unlike others. My eyes glance at expensive houses and street signs. The memories play. Walks with a friend, a crush, or a lover unravel. If the Kumiho is alone for the rest of her immortal life, is that sad? My eyes stare at street signs, remembering the path to the water, the cliffs, and the sun. What's so wrong with being by yourself? Is there a need for romance anymore?

As I pass familiar houses, I remember text messages, sunsets, and moments I had with others. In those memories, I wasn't alone. In the past, these people meant something to me. I imagine that when the

Kumiho is digging up graves for hearts and livers in folklore, she's surrounded by ghosts and souls, just like her. They are her company, ephemeral beings chasing misery away.

When I finally arrive at Sunset Cliffs, I sit down on the sand. I pull down my mask and sip at my turmeric matcha latte. Watching the waves crash onto rocks, I overhear conversations, smell the water, and gaze at the passing sailboats.

I might not be a Kumiho, but I'd like to dig up graves as the ghouls watch. I'd like to share a piece of a human's heart with the phantoms and enjoy their company. We'd say our goodbyes, then move on to another journey, wandering the world, chasing after fragments.

ACKNOWLEDGEMENTS

I would like to start off by thanking all the boys who made it into this book. If it wasn't for the messes we made together, I wouldn't have this book (though I wouldn't be as hurt).

Speaking of heartbreak, thanks to Joshua George, who has helped me through that misery. From receiving an (outdated) list of codenames with one sentence summaries to listening in kindness and generosity about each and every one of them, my life would have been much more difficult without your friendship and you by my side. Till death by *Rush Hour* events do us part.

To Ariel Advincula and Dannis Thompson, two of my best friends who took a road trip to Portland with me and then showed up to my birthday party to be sapped, you two have been there for me to offer guidance, support, and love through the hilarity and seriousness of my misadventures. You helped me turn them into written reflections.

Cheers to my parents for birthing me. Without your love, I wouldn't have examples to follow. So thank you for showing me what love is every day. Though, honestly, Dad, you know way too much about my love life and you slander me whenever you have a chance.

Slandering me in games, always, thank you to Alaric López, who played *Hollow Knight* via Discord while I was reading about open-heart surgery. You have been a listening ear throughout the book-writing process. Thanks for sharing time with me while my brain fried.

To Christina Vega, from sitting in a spoken word poetry class to Abby E. Murray introducing us via Facebook to being friends with Jonah, this has been a rollercoaster ride that has led us here, to this moment. I am so grateful and appreciative that you took a chance with this manuscript.

To Kate Threat, who has been a part of this journey from the beginning, your feedback, work, and presence have been a wonderful light.

The gorgeous cover art is all thanks to Kira Jacobson! I knew the moment Kira sent me an email about wanting to illustrate a different project in class that I'd consider working with Kira on this book. Your work inspires me and brings so much joy. To collaborate with you for a project is a dream come true.

From settling debates between Christina and me to talking social media, Jonah Barrett, you are an expert in all modern technology. Thank you so much for being a part of this, and I am extremely grateful to have had your eyes on and thoughts about all things virtual.

To Sam Hill, thank you for the brilliant layout design! All the work you've done has made the book's organs look so pretty.

To my mentors, both new and old, but specifically, Carolyn Gilman, Lily Hoang, Jac Jemc, Ever Jones, and Abby Murray, thank you for your teaching, guidance, advice, and support throughout the years. Thank you for encouraging me to keep going with the writing, especially about my love life shitshow.

To the MFA in creative writing gang, who have read early drafts in workshops, supported me throughout the journey, and even wrote reviews, thank you. Writing during a pandemic was painful, but you all made it so much more tolerable.

And last but not least, thank you to *you*, the reader! Thanks for taking a chance on this tiny little memoir. It's messy, crude, and sexy, and that's not for everyone, so thank you for reading.

ABOUT THE AUTHOR

Born and raised in Olympia, Washington, Alissa Tu is a Vietnamese American writer with work published in *The Margins* (Asian American Writers' Workshop), *diaCRITICs*, and *Honey Literary*. She earned her MFA in creative writing at the University of California–San Diego. Fresh out of messy romance gossip, she's currently crafting the perfect Vietnamese love letter.

ABOUT THE PRESS

Blue Cactus Press is an independent and hybrid publisher. Our mission it to craft books that serve as community resources for and by historically marginalized groups. We center the work of historically marginalized communities in our authorship, staff, and collaborations. Many of our books are about caregiving, community building, equity, and self-actualization.

Our vision is to foster heritages of storytelling in which voices and stories of historically marginalized people are visible, listened to, celebrated, and recognized as integral to the fabric of our communities. We want readers to believe their stories are valuable, and to see themselves and their experiences reflected in our books.

To support the press, become a member at patreon.com/bluecactuspress, and purchase books at bluecactuspress.com.

Confessions of a Modern-Day Kumiho
by Alissa Tu

Copyright © 2022 Alissa Tu
All rights reserved.

ISBN: 9781736820933
First Edition

Cover art by Kira Jacobson
Cover design and layout by Knic Pfost
Editing by Kate Threat, Christina Vega,
and Nisha Bolsey

Blue Cactus Press | Tacoma, Washington
bluecactuspress.com

www.ingramcontent.com/pod-product-compliance
Lightning Source LLC
Chambersburg PA
CBHW042130100526
44587CB00026B/4236